AF412659

ONE MAN
AND HIS SEA

. . . scuffed leather shoes.

GORDON S. SMITH

ONE
MAN
AND
HIS
SEA

*With photographs
by the author*

Hastings House,
Publishers
New York

Library of Congress Cataloging in Publication Data

Smith, Gordon S.   One man and his sea.
    1. Lobster fisheries—Massachusetts—Cape Cod.
2. Fishermen—Massachusetts—Biography. 3. Bloomer,
Harvey W. 4. Chatham, Mass. I. Title.
SH380.2.U6S64   639'.54'10924 [B]   78–17131
ISBN 0–8038–5391–2

Published simultaneously in Canada by
Saunders of Toronto, Ltd., Don Mills, Ontario

Printed in the United States of America

To Herself
who brightens my day
and lightens my load.

To myself I seem to have been only like
a boy playing on the sea-shore, and
diverting myself in now and then finding
a smoother pebble or a prettier shell
than ordinary, whilst the great ocean
of truth lay all undiscovered before me.

*(Sir Isaac Newton: quoted in Brewster's
Memoirs of Newton II. xxvii.)*

PREFACE

DURING THE EARLY 1970's there was much hullabaloo over the plight of lobstermen: foreign fishing vessels destroyed off-shore gear; American draggers ploughed through pots set inshore. The wreckers did not give a damn. Within the turmoil I believed there were stories of human conflict not reaching the public; stories of human emotion and changing traditions. Lobstering is an industry, yet the trapping of lobsters is as individual as a sand bar is to a rocky cove. I live on Cape Cod, so I commenced to search and cull for a co-operative lobsterman who trapped in the shifting sands off these shores.

I know a man who can steer me in the direction of fishermen. I went to him first.

Willard Nickerson can be found most days in summer operating his fish market by the Chatham Fish Pier. A good time for asking questions and getting answers, candid and without embarrassment, is before the summer folk come slamming the screen door and piling in to the glass topped counter. Eight o'clock in the morning is a fair time; too early for the onslaught.

Then he will be found cutting fish, and it would be prudent for anyone to stay on the public side of the counter when he is cutting, even if there exists the privilege to enter by the employee's back door. His knives are sharp and if he should be crowded his nerves tighten unbelievably, and questions don't get answered.

I said my piece from over the counter-top. He puckered his brow and pressed his lips together, but spoke no word until he had finished slicing. Then he took a scrap of paper and a pencil stub and began scribbling names. He wrote six. He said that he had more should I need them.

He read the list aloud, explaining that he might have misspelled a name and did not want to be misunderstood for that.

He underscored two and said, "You might get along with either of those, but the first runs a spotless boat." He raised his eyes and cocked his head with a poker-face that could have said anything. Then he leaned forward slightly over the counter and said. "His boat doesn't smell of stinking fish. You'll find his shanty by Mitchell River. Do you know where that is?" I said that I did, and thanked him.

Harvey Bloomer was the man. Lilly B was the boat. The day I went looking for him I found him standing by an old green Ford truck on his quay. There were two men talking earnestly together and I surmised one had to be Harvey Bloomer. I waited across Bridge Street to find out who would leave.

The man who remained was cleanshaven. He had a crop of dark hair. He was tall; six feet something. He wore a checkered shirt with the sleeves rolled above the elbows, plain salt-hay-green pants, and scuffed leather shoes. He wore salt encrusted bifocal glasses. His eyes were blue. He was lean. His frame had the structure of substance.

"Would you be Harvey Bloomer?"

"Yes, I am."

We shook hands and talked for two hours as if he had nothing else to do. It seemed to be his "do-nothing-day," one so flavored by Wyman Richardson in the book *The House on Nauset Marsh*.

Problems with foreign fishing fleets and draggers seemed not to enter the conversation. The man was conscious of them, knew many in detail as he was Vice President of the Massachusetts Lobstermen's Association. Instead, there developed the beginnings of an understanding for a way of living many of us have lost touch with since the Pilgrims landed in Massachusetts. This man, I learned, reaps a harvest from the sea, not for riches in dollars,

but for riches from work where work is the essence of living. He is the last of a three-generation-line of Cape Cod fishermen; a real Cape Codder. He has no children; no one to pass along the essence. What he knows and feels about his living will end with him. His story is not head-line news, but it is a story that everyone owns in different forms deep down inside, suppressed, sat upon by the pressures of his own living.

I had to change my course, to listen to music from a wind in a different direction. Instead of sleuthing for troubles on the seas of the east coast I took what confronted me. I followed those things that are ever slipping through our sophisticated selves and committed an inadequate few to paper, for how could I cover so long a human breeding in a book one-half-inch thin? It is but an essay, ONE MAN AND HIS SEA.

CHAPTER ONE

THE WEATHER REPORT was evil. For two days he had stayed ashore. Not because of necessity; it was in him to stay because it was his want. There were old pots to repair and new warps to make, and there was no special reason for the tasks except the excuse that he was a Cape Codder. Now there were pots off Monomoy Point needing attention.

The man finished his evening's bourbon and knocked out the ash from his pipe. Winds of twenty knots were all right by him if they had to be, but spreads of ten to twenty make for a slow, difficult trip. His wife had heard the report, too. She knew it meant difficulties hauling.

"Good night, Lil." He turned towards the stairs and his bed.

This man, Harvey W. Bloomer, is a fisherman out of Chatham, as was his father, George Washington Bloomer, Jr., and his grandfather, George Washington Bloomer. Whether his great-grandfather, William Raymond Bloomer, fished is uncertain. None attended a school of fishing. Knowledge was acquired by being where knowledge was. Learning was by watching closely, then by doing. Learning to splice a line was by being shown, but once. Learning to navigate was by being near the compass and observing the signs in the environment. Many a time his father had said, "Take over, Son." He proved himself before his teens. Apprenticeship it was called; a name for sound schooling.

. . . only a dory to row in.

Harvey Bloomer had had his schooling. His tutors had passed on. Except for the knowledge and the passing on of it between generations he owes nothing. He is independent. He holds the seeds of his living. He is a capable man, living off the sea by himself and preferring the living that way. In his youth he had made the choice to be a fisherman, although he was not without the opportunity to travel another course. He had made his choice and he lives by it. He would say, "A fisherman shouldn't starve. There is plenty of getting from the sea if he has a mind to it." There is plenty to reap, and Harvey Bloomer is schooled for that thinking.

His grandfather's fishing was no special chore. He had fewer pots and only a dory to row in. That was enough. In those days fishermen summered with their families on the Point, the extremity of Monomoy Island. They had the advantage of being 45 miles out to sea from the Massachusetts mainland and 10 miles out from Chatham. Today, the Point is a Federal Wilderness and no fisherman may live there.

Grandfather Bloomer put out at dawn, into a stillness, into a ubiquitous white fog, onto a calm surface, over treacherous bottoms. It was into a calmness that went with the sound of oars dipping quietly under a gentle rhythm. The blades would eddy the water. On the shorter return strokes pearly droplets would trickle back into the virgin swirls. The trip was with the dory slicing at the bow, heading to unseen catches. Guidance was by cognition to the beginning of the set in a mysterious vastness where Nantucket Sound and the Atlantic Ocean meet. The hand-hauling of pots was to nature's voices; a calling gull kee-ow-ing from somewhere to his kind in a nothingness place. It was buoys of white cedar, warp of manila line, lobster plugs from hand-whittled basswood. It was expectation; a curse at cod flapping instead. The fog would burn off. A few moments were fulfilled with the oars at rest, and a tobacco pipe lit.

Harvey Bloomer had this nostalgic knowledge tucked inside him, but he needed fast sleep. His generation can not rest on oars. He has more catching to do. He has farther to travel (10 miles to the Point to meet the dawn), more pots to tend, more equipment to maintain, more overheads to spoil his financial returns, and lobsters less plentiful. The commuting to and from the fishing grounds consumes precious time, but a fast boat helps, when the weather permits the speed.

Lilly Bloomer wound her wristwatch and turned off the living room light.

CHAPTER TWO

THE MAN'S HOME on Barcliff Avenue has a high elevation. It was built as a sea captain's: back from the ocean; away from flooding hurricanes. Yet the wooden dwelling stands only half-a-mile from the Chatham Fish Pier. The ocean is both distant and close. To find a view of it from a second floor bedroom window a person has to hunt through a foreground of trees and roof tops. The home is well located for a man who knows the ugliest elements and who must have a close intimacy with them.

There is about the man an aura that embraces these elements in time and space. He checks weather forecasts, but meditates upon how far wrong they might be. Forecasts are not true to him; only an official opinion. His weather-thinking, as well as his lobster-movement surveillance, surfaces at any time. The two are part of him as if he were keeper of both.

The sun was an hour away when he stirred between the sheets.

He travelled Barcliff Avenue that runs easterly to Shore Road. The junction is close to Chatham Fish Pier, but he felt the sea's presence only slightly. Its major awe is seen at the bluff by the lighthouse. There the business of the sea and the human relationship with it appears in sanitary coldness: the white Coast Guard station; the white Coast Guard's amphibious duck; the white Chatham Light.

At another time the ocean view might warm the soul: the rising sun searching for a break in clouds and mist; a lone freighter from Boston heading south, looking submerged with its heavy cargo; three draggers seeking company out there.

Off the back side of Monomoy the blunt dark jagged end of the Pendleton wreck sticks up as a jolting reminder of the past. Three-thousand other disasters are not easily seen.

A low hedge cut to the shape of an anchor complements an obelisk next to the lighthouse. An inscription to the lost William H. Mack, the barge Wadena, their crews, and life-savers, is there to tell if tourists should turn around from the ocean and read.

> "Twilight and evening bell
> And after that the dark!
> And may there be no sadness
> Of farewell
> When I embark:
> For tho' from out our bourne
> Of time and place
> The flood may bear me far.
> I hope to see my Pilot face to face
> When I have crost the bar."

At low tide, Chatham Harbor, a channel running between the Town and the Outer Beach, is streaked with sand spits. Navigation markers keel to the tides. Yellow-legs, sandpipers, and sanderlings tread the shallows. Common and roseate terns fly in ritual maneuvers. Herring gulls drop shellfish on packed sand. Canada geese migrate in straggly streams, flying in the wrong direction and knowing why.

Except for the human touch, the scene is a hundred seasons old. Sand bars and beaches are always shifting, always changing. The sands move south, ever extending the contours of shoals, ever re-writing the sea bottom faster than the skills of chart-makers or buoy-placers. Knowledge of this bottom is important to this man's living. Placement of canyons and wrecks, movement of tides, temperatures of the water, weather, all are read to make him a top-liner.

His old green dew-covered truck took the final rise of Shore Road to the bluff and Chatham Light flooded his path. Spurious light tipped the town's roof tops, and a cusp splashed momentarily on a church spire. Ghostly shadows fleeted repeatedly along deserted streets—Silver Leaf Avenue, Water Street—drawing attention to Bridge Street, a straight road with a drawbridge.

The morning had its usual quietness. The grays were fading. Quietness here was not of the Atlantic. Out there was a running sea.

He gunned the gas pedal and the truck belted down the straight road. Just before the hump of the wooden bridge he braked, automatically took in Lilly B at her mooring, the skiff, the wharf, the property. He subconsciously scanned the scene for completeness. Stage Harbor Marine Service was in darkness. The lights along the front of the marina still glowed, giving way to a morning in the east. He was first hereabouts, although a skunk and two muskrats flinched to the routine rumble of loose bridge timbers, the dawning signal of human annoyances.

Up river is Mill Pond. Out of sight and almost into town is Little Mill Pond, a tidal water far back from the sea. Close to the wall of the quay cruised a sand shark, gracefully, lissomely agile, silver-edged, a potential swimming power house. It came quietly and silently vanished. Along these same edges at low tide men will stand thigh deep digging for quahogs.

This is Mitchell River. The tides sweep in here. Down river, to the right, is the route to the sea, through Stage Harbor and a cut in Harding Beach.

The man's place is across the bridge. It is a jut of raised land, a fraction of an acre, reclaimed from salt marshes a generation ago.

Two gray shingled buildings stand solemnly on the quay. The larger is the store shanty. The smaller, the bait shanty. From the rock-lined wall of the quay is a sturdy wooden T-shaped wharf. To one side is a macadam ramp. The rest of the quay is fill, with a sparse topping of crushed clam shells that can reflect a searing brilliant sun.

. . . not a pot or buoy is seen.

Depending upon the season, so rests the paraphernalia around the shanties. In winter it is cleaned up. The skiff, which is normally in the water ready for transporting bait and lobsters between the ramp and Lilly B, would be upturned by the store shanty. Perhaps five old discarded lobster pots would remain leaning against the bait shanty.

There would not be much else to see in winter. Lilly B would be high and dry under cover at the Old Mill Boat Yard.

When snow is flecked around, the herring gulls, who call the place home in summer, will be elsewhere scrounging food, and the man and his wife will be in Florida. There is no lobstering from here then.

In summer, the place reflects its industry and takes on new meaning, and those who know the man can read in these things the goings and comings of him.

In spring and fall his place brims with pots; matured pots and new pots. These are the seasons when he sets and returns gear.

At the peak of lobstering not a pot or buoy is seen, unless it is a lost pot that has been found again, and now encrusted with barnacles and trash. But around the store shanty, fish boxes pile high. Metal floats and bric-a-brac found floating in the course of hauling are dropped at random. Giant conches are dumped into containers, and herring gulls periodically till for left-overs. An art group will invade the quay and wonder at the business of the man as he reluctantly asks permission to go about his work. The intrusion breaks his rhythm, a hard working, steel swinging rhythm. He earns his buck.

As he turned into his property this morning a police cruiser patrolled Bridge Street. By the time the cruiser rumbled across the bridge there was no visible sign of him, only noise from scuffed leather shoes on crushed clam shells.

CHAPTER THREE

THE CRUNCHING gave way to quietness from boots on marginal sod as the man approached a rusty stake at the quay's edge.

Mosquitoes walked the wharf's bollards searching for splashed blood left from the spring's cod fishing. They did not bother the man and he seemed not to notice them. But he smelled the smell of running bait. He had sniffed the air and found the oiliness that was rising to the surface in the untidy eating from the chasing bass. He knew the struggles below the surface as he knew the patrolling police cruiser had stopped where Bridge Street meets Stage Harbor Road. But he did not rationalize with the events. They were of no consequence at the moment. He was more acutely conscious of things ten miles ahead of him.

While aware of the early morning's happenings he worked through his chores.

He untied the skiff's line from the rusty metal stake and began hauling, his body swaying backward and forward under alternate strokes. His head high, his eyes were in abstract gaze. The bow of the skiff gently touched the ramp. He retied the lines to the stake with clove hitches and untied the painter from the line. Taking up his lunch box he stepped into the skiff, placed the box on a thwart, picked up an oar and rowed paddle fashion in a standing position. It was a short distance. He paddled from

side to side to steer her and came alongside Lilly B, climbed
aboard and bent the skiff's painter to the cuddy post.

The process was as automatic as the firing of spark plugs
in Lilly B's engine; as automatic and in sequence as it had been
three days before, and the year before that. For these are habits
and movements that relieve the mind from conscious things, from
things unpredictable, from things out of a man's immediate control. For this man lobsters alone, and habit is a good routine. It
relaxes the body and mind in readiness for contemporary emer-
gencies.

He unlocked the cabin and started the engine. It started at
the first touch of the button. The engine sounded good. It throb-
bed that low throated sound, one that resonated below decks.
The man revved the engine once, and let it idle; a human trait,
and in this respect the man was as other men. But the engine
did sound good.

Then he fixed the davit by swinging out metal arms and
bolting them together, while the boat was steady and not rolling
out in the Sound. Then he switched on the navigation lights.

Returning to the skiff he boarded her once again, slipped
the painter from the cuddy post and pulled himself along to Lilly
B's bow and mooring line. He hauled up some of the line's slack
and exposed a large ring that he slipped over the skiff's stem.
Now the two vessels were moored to the same line temporarily.
He climbed Lilly B's bow to her foredeck, slipped the mooring
line from the bitt and dumped the bitter end into the skiff. Lilly
B was free. He scuttled back to the cuddy and cautiously piloted
through the channel.

As he navigated by the moored small craft of the river he
scanned the approaches to Champlain Road for activity, for headlights from a truck. He checked for a lobsterman, "Sonny" Mallowes.

Sonny was kind of special to the man. Sonny was the antithe-
sis of all the man did, which seemed to be a reason for making
him special. This man was successful; Sonny struggled for a living.
This man had a clean smart craft that almost smelled of shaving
cream and hair oil; Sonny's boat was somewhat unkempt, smelling

of stinking fish. This man's craft was in top working order; Sonny's was half-working: oil from the hydraulic hauling winch might spill regularly onto the deck. This man changed spark plugs at the slightest hint of failure, because failure meant lost time in the run, and the catch might spoil because of it; Sonny's 6 cylinder Ford diesel was comparatively slow.

This man's business was lobstering. Lobstering would be his way of life for as long as he could lobster-fish. He knew he was a top-liner. He had success as no other lobsterman fishing these waters could ever have. Sonny would end the season wondering whether this should be his last. He might give up the trade for something else next year. There was no money in this rotten business. The two men fished the same waters.

The man helped Sonny along as best he could, more like he might a son if he had one. He had Sonny place pots alongside his own. He would give him tips that many another lobsterman would guard as family secrets. Yet Sonny still struggled. Success and struggle ran parallel; so typical of the ways of fishing in general and lobstering in particular: everything similar, each man having similar opportunities. One profits, the other suffers.

The man was now into the cut by the Old Harding Beach Lighthouse when he saw headlights swing around a bend in Champlain Road. He thought that might be Sonny, and late. He wasn't sure that he was happy if it should be him, the way the seas would be running.

CHAPTER FOUR

MITCHELL RIVER and Stage Harbor regained their original solitude. Except for the absence of Lilly B, the changed position of the skiff, the truck parked between the two shanties, things were again much as they had been. The morning's coming light spread higher. The headlights that had picked their way along Champlain Road penetrated the suspended dampness in Port Fortune Lane where the old freezer plant once stood before the fire. From the truck's cab Sonny watched Lilly B disappear.

The town of Chatham is like any other north-east New England coastal community at this time of morning: slightly stirring. The Mobil gas station at the rotary where Main Street spins off Route 28 had lights blazing, a sign of a restless proprietor. To a casual visitor who wanted to know the feel of the place, and to rise this early to seek out the hidden activities, he would be disappointed, for the business is not planned as in an inland city. Weather, seasons, moons, natural and unnatural vagaries of the catch are more dictatorial here than the plans of man. Bridge Street would certainly seem dead; without the knowing the visitor couldn't read the signs. The fishermen were not lazy, but most of Chatham's fishing fleet would be moored in Aunt Lydia's Cove, perhaps 35 boats all pointing south-westerly with the incoming tide, and occasionally veering to port with each north-easterly gust. The weather and Chatham Bars would keep them in.

Yet one man held his shoulder to the yoke, and not for pride of Town or casual visitor. It was in him to reach his pots that were overdue for hauling; overdue for a three-day set. Ugly winds or no, he had put his mind upon the trip. He had his shoals to cross, rip tides to watch, on his Nantucket Sound side of Monomoy Island. Chatham Bars was not his course.

In the speed zone of the channel the headway was irksome. Later, out in the Sound, the forceful drumming from his 365 horse-power engine would better satisfy his eagerness—at 13 knots if the seas would allow.

Lilly B bore down on the white beacon at the end of Harding's Beach and the man commenced to change course. Morris Island withdrew into the grayness. As he changed, a dark mass on the dike slithered into the cut. The man saw the irregular form disintegrate and recognized the eeriness as "hair seals." They moved clear of the man and his noise. In the dimness the species could have been the Atlantic seal, *Halichoreorus grypus.* They breed as a small colony on isolated sand spits off Muskeget Island in the Sound, but the harbor seal, *Phoca vitulina,* is common. Phocidae are often referred to as hair seals in the trade. Man and seals moved away from each other; both observing, the seals suspicious and cautious.

The man's course had swung south-westerly as the low tide had him skirt the South'ard, a tidal flat west of Monomoy too shallow for his craft's three-foot draft.

Hardly was he out of the cut when danger straddled his course. A drifting, barkless, tree trunk lolled in the troughs like a petrified sea serpent ready to tear into Lilly B's hull.

He remembered a similar morning when he was far out into the Sound making time in the darkness to his pots when he rammed into the butt end of a tree trunk and holed "Smallfry," his smaller craft of that time. Water gushed in. Too fast for the bilge pump. There was no contest with sea water entering a hole the size of a fist. He gave the alarm and bailed madly with a bucket. The Coast Guard hurried in and dropped two pumps.

For a moment the incident quickened his pulse. He steered away eyeing the danger as it cleared his stern.

They breed . . . off Muskeget Island.

The man knows the dangers of working alone. He also knows the perils, particularly to elderly men with complacency. He recalled fishing with his father and catching him at the wheel, dozing, and how he had to think of a subtle excuse to bring him to. There can be no liberties. In working alone there is no back-up for escape. The man's subconscious says, "I'll never go to sleep on my boat."

Yet he complains gently. He knows the changes in him as he grows older, and he seems to understand the process better than most. His life's run, if heredity has sense enough to prove itself, will see the end of the century. He knows the biological processes are working just as they are working on lobsters. He knows that his reflexes are slower than when he shared the load with his father. He battles to keep acute his senses: of sight, and touch, and hearing, and balance, and smell, which are kept surprisingly attuned to the subtleties around him.

Any other morning there would be moving navigation lights from sports fishermen in Boston Whalers leaving Wychmere, Saquatucket or Allen harbors. On the man's return trip, these same sportsmen would come alongside with false confidence. They would speed ahead. The man would muse the moment and weigh the craft and ease forward the throttle. Lilly B's semi-planing hull would skim the wave crests. Barely touching the wheel, the man would stand in the cuddy shifting his weight from one foot to the other, chuckling, smiling, and waving a hand. But he would not speed for long. Just enough to prove his point. He had too much respect for his engine for that foolery. This morning there were no lights.

If there was any sleep left in the man it was soon to be shaken out of him with the pitching and tossing and rolling. He constantly pivoted his head around like a human owl, looking, checking, satisfying himself that he cruised within visible safety. He scribed an arc around himself, observing the confines like the raster on a 'scope. Chatham Light was not yet visible. Obscured Morris Island blocked the line of sight. But he checked.

When the light eventually slipped out from behind, he changed course again, a degree to port. Flash . . flash

. . . . flash . . flash . . . . a group flash of two white
lights every 10 seconds. At an elevation of 80 feet it was officially
possible to see it from 15 miles off. The man hauled pots just
that far away, but in sea mist.

Having seen the light he stood like an ice-skater performing
the spread-eagle, eyeing Lilly B and her wake for alignment and
course, practically disregarding the compass. If his eyes flicked
in that direction it was but for a casual moment. He listened
more to the engine, now hot, and used the changing tone as
some mysterious message from the sea bottom; detecting the shallows and the depths by the note. If there was a change, it wasn't
apparent. The engine roared. But if one checked the engine revolutions, a change was there; a small matter of 100 a minute between
the depths; faster over the shallows.

Salt spray continually doused the bow and cuddy and his
vision. He cracked open the windshield and peered through the
narrow slit, much as a gun officer in a bunker. The slit offered
a cleaner view, but not the same protection. Dampness and drying
salt smeared his bifocals. To clean his glasses would have been
futile, so he adjusted to the condition and was immune to the
handicap.

Unseen, but not unknown to him, were weir traps a mile
to port. Something not to get enmeshed in. Something to reckon
with, particularly at high tide if he had chosen to take the course
over the tidal flats. Pleasure boatmen curse the weirs, but part
of this man's livelihood depends upon them. From those who
fish the weirs he buys skate. Brined skate was slightly nauseating
in the tubs on deck right now. The sickliness mixed with heat
and fumes from the exhaust swept over the stern. And his is a
clean boat!

The engine's exhaust pulsed through twin pipes amidships.
They turned upward behind the engine and extended to the trail-
ing edge of the cuddy roof like two thin fingers, except that
these had arthritic swellings for mufflers. They were exceedingly
hot. Heat radiated to the man and he could feel it upon his face,

When the light eventually slipped out . . .

[27]

. . . weir traps a mile to port.

and by its intensity he could have told this morning's distance run. But where he might be pitched onto hot metal he had asbestos wrapping. In the spring and fall when he hand-lines over wrecks for cod his cuddy is partitioned all around and the twin pipes, partially inside, keep him warm. But now the warmth wasn't keeping peace with anything, and certainly not with the summer humidity.

The deep sea trough suddenly opened under the bow before the man was ready. Lilly B took the pitch and nose-dived. The man lurched onto the wheel and grasped the throttle in one continuous scurry. "Christ!" he said to the wind, and reduced speed. Being unto himself, the seas, and the tossing, it was his prerogative to blaspheme and spill the sense of it overboard to whatever and whoever might be there. Because of that his blood pulse quickened.

These were not high seas. Monomoy Island is a barrier sand reef giving a lee in the Sound for a north-east blow. Beyond Monomoy Point is open water to the Atlantic.

From the shore to beyond the Point, amongst the shoals and rips to his first set, would take an hour; a commuter's run, eating into time with nothing to show; day after day, weeks, months; commuting year after year.

A commuter's run, not on Long Island's railroad to New York City, but a necessary run. Not in smoke-filled compartments with coughings and splutterings from jostling flesh. Not dependent upon engineers at controls; sitting, waiting, reading the financial news, a stranger as companion, conditioned to this society. Not being swept like a pawn in the moves of a corporation's shadow to Los Angeles, and sitting lonely in a traffic jam, on a different run with different strangers peering from behind steering wheels, knowing nothing more than the points of departure and arrival.

His is a commuter's run forced upon him, but forced because of the changing habits of lobsters and lobster scarcity; over fishing; demand from human population increase. There is no more in-shore lobstering off Chatham that amounts to a living. So he commutes. To the sound of the wind, to the salt spray flying,

. . . to the sight of Canada geese.

to the buffeting from waves, to the call of passing herring gulls
and diving terns, to the sight of Canada geese, to the raft of
black ducks off the bow, to the sea mists moving, and to far
off transports cruising. It is to the necessity of the overall changing
environment.

Rodgers Shoal: 26 feet, 23 feet, 14 feet, 10 feet, 12 feet, 4
feet, 4 feet, 4 feet . . . the water boiled. Monomoy Point, the
outline of land, low, sandy, silent, empty of human life. By habit,
the man navigated close to, and the water calmed. He crossed
the 40-foot deep channel and eased Lilly B a few degrees. Again
the water boiled in madness. Red bell buoy number 10 sliced

into the incoming tide and drove a wedge into her with a 200-foot wake. The intimacy of the Atlantic bore upon the man and his boat. The water chopped and turned light green. There was talking with the bottom and the surface and the man eased to the east into it all . . .

Fathomless. Gray, deeper, deeper. A leaving the last edges of finite land. An entering into a rolling world of seclusion; an expectation of incidents that stayed out of reach; a beyond that outperformed the bounderings of a closing mist; a universe that didn't hold to city experience; a dream-scene of truth. Butler Hole: 117 feet.

A movement that reflected the elements; a motion in Lilly B without knowledge of travel. She pitched and rolled and plowed, and there was with her a wake, too, but there was no more apparent going forward.

This was the entering of the place.

This was the entering of the place.

CHAPTER FIVE

THE SEA CONCEIVED the idea and the place with the receding Ice Age. After conceiving, there was molding, shaping, reshaping, the taking away, the adding to. Nothing was final.

Off Monomoy Point grew an overwhelmingly large blue mussel bed, until draggers of the 20th Century scoured the sea bottom and predator starfish sucked their way through with their inside-out stomachs and the seas rolled the bed up like carpet and carried it far off in small pieces. The bed had been several miles across in each direction.

Kelp grew, too, in large fields. Fish came and were everywhere over the bottom, feeding on the growth that had worked its way in after the big ice retreat—and smaller fish, crabs, and worms. Fishermen came and anchored to this bottom and pulled in their lines for the abundance that had been attracted. And the fishermen found the fishing good. They caught market cod and small steak cod in fine quantity. Some caught mackerel. Those that trapped for lobster did well—two cents apiece for the big ones; for the others, one cent.

This was before this man's time.

By the time of his grandfather lobsters were 14 cents apiece for any size; one clawed lobsters were two for 14 cents. The fishermen knew the waters, knew the treacheries of the fast currents; 3½ knots that were dangers for the unwary. This man knew them, too, in his time.

The unwary came in thousands. Three-thousand wrecks have sunk into the Cape Cod sands. Off the shores of Chatham the wooden tombs account for half that number, so it is said. Miles off the shores, passed soundings of 120 feet, the bottom again makes love to the surface; they kiss together at 4 feet at the Rose and Crown. The breakers are heavy. And the destiny of America might not have been great if the Mayflower had not turned from Pollock Rip.

The roster of ghosts reads like a scroll to the World Wars' dead:

Shovelful Shoal
schooner John C. Smith—1904
schooner Alexandria—1887
schooner Bessie Parker—1895
schooner Ocean Traveller—1876
schooner M. M. Merriman—1891
    etc.
Handkerchief Shoal
brig Sallie—1852
bark H. C. Sibles—1886
schooner Sarah Potter—1903
    etc.
Stonehorse Shoal
schooner Thomas D. Harrison—1882
schooner Frederick Fish—1878
schooner Henry D. May—1913
    etc.

The character of the bottom is conversation for local tongue: Bearse Shoal, Twelvefoot Shoal, Little Round Shoal, Great Round Shoal, Orion Shoal, McBlair Shoal . . .

This man knows. Apart from Sonny, he is the one man who traps for lobsters here. He knows the sea bottom and reads it like Braille. He knows the tides and how they affect the depths and the currents over the shoals. He knows the movements of lobsters here, the way they travel, and where to sink pots. It was off the Point that he learned these things, when the family moved house each summer to fish from these shores, picking up the knowledge from father and grandfather and the others of his age who messed about with boats and fishing. There wasn't much time for mischief then, but if it was mischief, it was without hurt.

This was the place.

CHAPTER SIX

THIS MAN AND the place joined together because of circumstance: because of a fragile line in his ancestry; because of the seeds of evolution and support for lobsters. Each travelled to a common sanctuary. To another man the premise might be filled with unsolved questions, or be a wasteland in the sea of oceans. To this man it is as familiar as finding his way to the bathroom at two o'clock in the morning. Without the bed as reference he might be lost in his own home. With the shore as reference he knows his orientation from the point of departure to wave top to wave top. He knows where he is going and where he came from—even in fog, which is not so much a curtain as an element of the place. Fog and sea mist are ever hangers-on. Whereas the Province Lands, the Barnstable Marshes, Woods Hole have their full day of sun, off Chatham, in and around Nantucket Shoals, fog clings close.

It is in him to know where he is at each and every turn. His navigation is a series of steps; a leaving from one known lodgment to the next.

There is reasoning in his progress, but there is much unreasoning, too; that is, if we try to reason for it. Gulls and terns roam air currents in storms and fogs and seem to know where they are headed. Fishes and creatures of the oceans seem to know something about their elevations and their whereabouts in their migrations. Lobsters seem to have senses that do not parallel those of man's, and man seems at a loss to understand them because he has not those senses to understand them with. Or so it seems.

There is much in man that is dormant for lack of necessity. With the need, with the ancestry, with the liking, that which is dormant unfolds. So it is with this man.

If he happens to have you aboard and after an hour's run in fog you ask, "Why are we changing course here?" he is apt to reply, "This is where we change."

If you persist with your questioning, he will tell you the compass bearing, the revolutions of the engine, the time since leaving shore, the amount of drift, and facts that might occur to him, implying that these can be juggled to the point where he had changed course. Perhaps these facts will make you happier in your understanding of things navigable. You, of course, trust him. You know he knows the facts, and you know the man knows in himself where he is because he is familiar with the place without the facts. You would have faith in him, and he has his own.

The man went into action. Off came his shoes. On went his thigh boots. He threw the tape of a yellow waterproof apron over his head and tied the waist strings around him. From a cardboard box in the cabin he grabbed a new pair of cotton gloves. He pushed his hands into them, swung around, selected the nearest of two five-gallon drums of bait and hauled it on top of the engine-box. Alongside he set three empty bushel baskets: one to hold a gaff and the retrieved buoys waiting their turn in the run; one for culls and large lobsters; one for selects. From the port side of the cuddy he grabbed a worn square pegboard, large enough to place a lobster on, and put it on the shelf above the engine controls. He also reached for two plastic ice-cream cartons filled with wooden pegs; sizes large and small.

In the coming light that was nearly ready the man switched on the depth recorder. It searched out the sea bottom. When the water is turbulent, as it was this day, the Bendix instrument would not faithfully relay its findings. Infinitesimal blue sparks seared a strip of paper with a facsimile of the ocean's bed. Sparks danced like miniature fairy lights, and there was a slight crackling for music at each blue burst. The man left the device operating, but he did not appear to depend upon it; his subconscious had finely tuned into his protoplasmic compass and the forces of Coriolis.

Automatically, he reached through a hole in the engine-box and engaged the clutch that would drive the hydraulic mechanism for the hauling sheave.

Again he turned his attention to the depth recorder, doodled with its adjustments, switched it off and coaxed Lilly B along as if the recorder was a waste of time and that he had better get back to the task of heading for the first buoy.

Daylight and arrival came together perfectly. It was as though the man had punched a time-clock for business. Stonehorse Shoal was under him. He had picked his way around to come against the elements, had come to the beginning of the set at a slight angle, the better to cross the line of buoys, if he had to, to outwit error. The elements were not directly against him, not head-on, so he would have to zig-zag a course all along the way, and he knew that the warps would pull up from under the hull instead of straight up, delaying him at each haul. His position seemed without orientation, a stationary universe, yet out of the void, mysteriously, beautifully, rocked with the surface wind, floated a buoy, first as a slight distant etching in the gray, then nearer and more defined, quietly working forward to starboard.

The man searched for the colorings, recognized that the red and white were his, and checked for the number and the set.

In the fogs and in paths of transports this trapper hauls for a minimum of six hours each day without pause, maneuvering over the canyons, the sand bars, the hidden timbers and ironware of wrecks, the gradients of slopes, the bottom compositions, gauging his luck and the movements of lobsters. His hands are salt-wet, saturated, and the pink from under the finger nails gone because of it, and one nail mutilated from a pugnacious crustacean, and a healing gash on a finger from another.

Now the monotonous rhythm of hauling began as Lilly B picked her way down the line of pots—a rugged slow three minutes to a pot as she rode over the warps. The man said to himself, "Just like a fisherman. Ya lay in awhile, then ya go when its rough."

CHAPTER SEVEN

THE STARBOARD BOW snugged upon the marker buoy at the precise approach. It was a coming together after a separation, a visitation in the morning's schedule. What of this rendezvous? Would it be a male or female keeper? A one-pound chick, a one-claw cull, a no-claw pistol? A lobster in the process of molting; a shedder? One that was undersize, a short; one that measured less than $3\frac{3}{16}$ inches from the inner part of the eye-socket to the end of the carapace? Perhaps a berried lobster; an egg-laying female? Would there be more than one, maybe two, three, four? Or no lobster? Would the pot be clogged with seaweed, mostly kelp? Would it have only a creature suitable for instant bait? A sea-robin? A cod? Would the bait be gone and the pot cleaned empty? Would the pot be pilfered? Pilfering is in fashion. But at rare moments a thief has donated dollars in a jar inside a pot as forgiveness. What of this rendezvous?

The man maneuvered at three knots, ready, poised with gaff in the right hand, nosing the wheel and engine controls with the left. He closed to eighteen inches, reached over the toe rail, bent his body into an inverted U, stretched across the freeboard and hooked the gaff neatly below the plastic buoy where the warp was tied to an eye made from an old rubber tire. His body straightened to 6' 2''. He pulled in slack, enough to loop over the idle pulley on the davit and around the driving sheave.

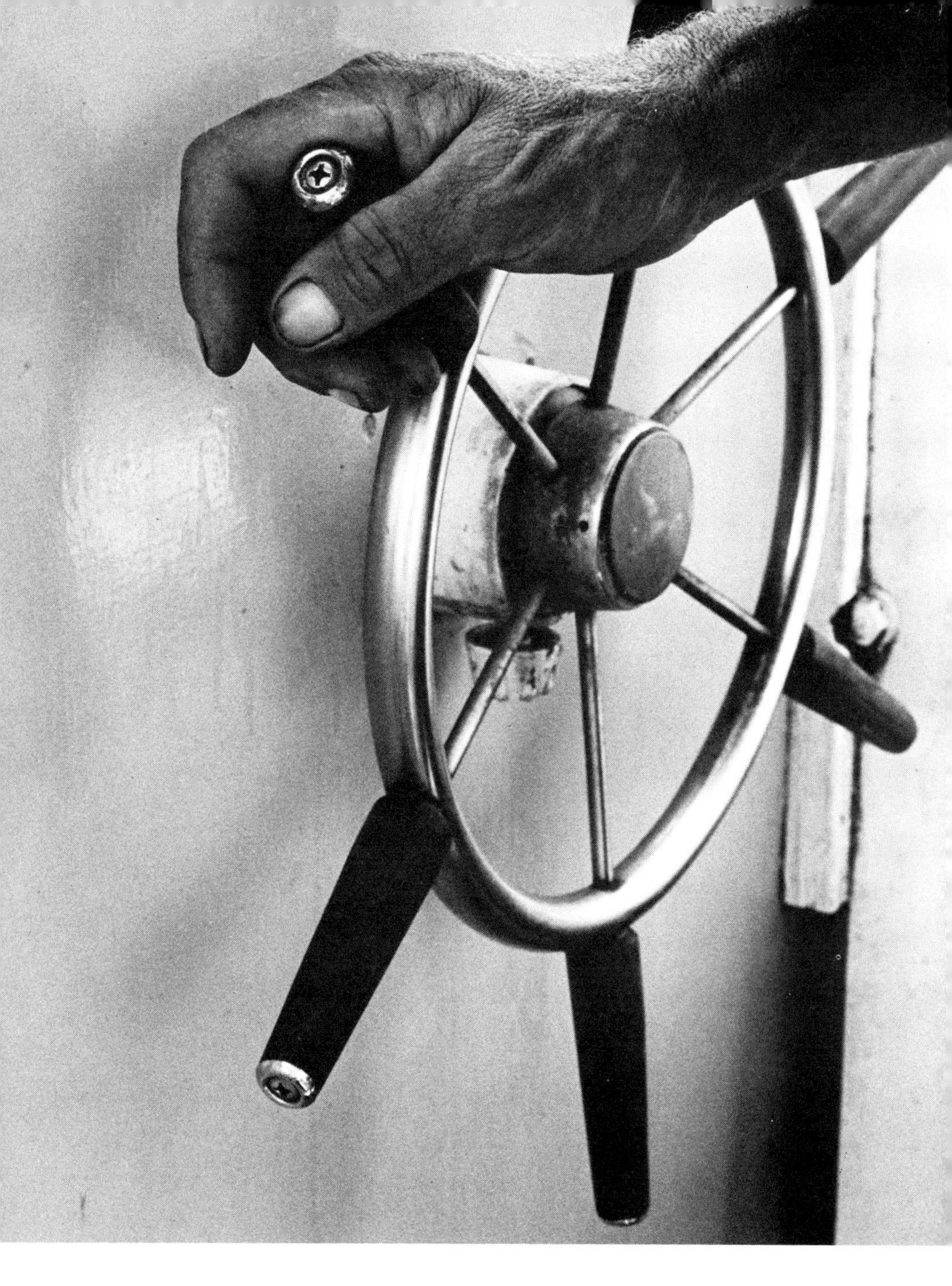

The man maneuvered at 3 knots, . . .

The rhythm and the play took but a few seconds, the body swinging and retrieving in harmony, an operation that might be a "rubber stamp" routine, yet as different from the next one coming, and the next, as one grain particle from another in a sand dune; a tune with deft changes.

The 20-fathom warp tightened like a banjo string. In the system the hydraulic fluid gurgled. The driving sheave moaned out a grinding that proved the tension. Two-dollars-worth of synthetic line rushed upwards.

Water within the lay of the line carried to the idle pulley. At the turn it spun off into the air. Where the warp immediately broke surface a triangular water-pane developed, shaped like the first dorsal fin of a basking shark, only transparent. As Lilly B tossed, it disappeared, to re-appear again on the next rise.

The warp looked thin and fragile and not suited to the load that had suddenly made it taut. Yet it was through the warp that the strain came about. It had only to part or give up the pot to release itself.

The man watched and listened. It was a practice that told how the hauling was coming: the watching for the drift of Lilly B and to get her stern away from the warp if he could; listening to the struggle between the warp and the drive—the wedging of the line into the V for bite.

A fixed metal finger two inches from the sheave periphery and 180 degrees around from where the line and sheave began to bite pushed the line away from the grip and spilled it onto the deck in slack irregular coils.

At ten fathoms a wafer buoy broke surface. It shot up and down the warp between pulley and water like a living creature happy to be coming aboard. With three fathoms to go the man eased the warp's tension. The wafer buoy dropped to the surface, and water that had been flowing along the warp like glazed ice in motion evaporated. Now the man could see the pot in the water, see the kelp waving. He knew that those long flat fronds oscillating through the slats, eel-like, would inhibit lobsters from the pot. It came through dripping.

A new presence was aboard; a sea bottom. As the pot twisted

around from its support the man turned it so that the door came about-face. He grabbed the lower end, the end with the head, pulled the half-round parlor pot over the gunwale onto a barrel that stood at coaming height and opened the door. Hermit crabs scurried in panic: small ones fell through the openings onto the deck and rolled over and over towards the scuppers at the stern; large ones were scooped out and thrown into the sea, along with empty waved whelk and moon snail shells. The latter are favorite mobile homes for hermits. The kelp was pulled off and also thrown back. Sand crabs hung on wherever they could. Others lost their grip, fell, and were crushed under 220 pounds of stomping boots. The mess that had started aboard would accumulate until shovelled away with a dustpan.

The living habits of hermit crabs increase the man's work load. They are attracted by the bait in the pots, where they change from one shell to a larger one as they grow. Their being scared out of their shells by lobsters caught in the pots causes stockpiling of empty shells. Their removal is a repeated chore. Each pot collects them and the accumulation slows the hauling and induces a house-cleaning that the man is resigned to.

It seemed that his attention was all with the pot, but he was aware of the drift and forward motion of Lilly B, and there would be time enough without fuss for re-alignment with his course when he had finished.

Not a scrap of bait was left on the center wooden spike. Bait consumed is dollars that must be taken into account, and when that bait is gone with no catch, that is a loss, complete with gasoline and time.

He took three pieces of skate that together weighed nearly a pound. One after the other he jammed them down hard onto the wooden spike. He closed the pot door by pushing against the tension of the old tire hinges and slipped the lower end into place. With abandon he slid the pot overboard. Eighty-five pounds of concrete ballast and saturated piss-oak hit the surface, breaking the seas into a billion faceted particles of white.

As the man wheeled Lilly B and made headway the warp ran out from the irregular coils on deck. At first it ran along the gunwale to the stern. Then it found itself a notch on the toe rail from previous wearings and gathered the rest of itself from the haphazard coiling and travelled along in a series of curves, like piglet tails twenty inches apart. At the last moment the man flipped the marker buoy overboard.

This was the meeting and the parting, and nothing to show for the encounter. The pot struck sea bottom by the time he came upon the second buoy. He guessed he might have better luck with this one; the first and last pots in a set occasionally do not fish well.

This pot broke surface with noise in it and he knew again that there were no lobsters. Fish and lobsters do not mix in pots. The cod flapped frantically. It was under 2½ pounds. The man called it scrod. He flipped it sideways to the deck towards the stern where it slithered on its side, flapping and gulping and seeking leverage. It lay there gasping, its body rising and falling faster than the heaving of the deck. There was no mutual motion as it slipped into death.

The third pot fished better; four lobsters, but only a five-pound keeper. The rest were two shorts and a two-pound female with a few brown eggs remaining on her. The general lobster laws prohibit keeping such a lobster, or having one in possession. But Chapter 130, Section 41 and 41A states, ". . . a person who takes any such lobster and immediately returns it alive to the waters from which it was taken shall not be subject to such penalty." Yet it is common practice to leave a small (under 3 pounds) brown egger in the pot until she spawns if she is not too full of eggs.

Unscrupulous lobstermen will take the berried lobster and remove the eggs. Even when performed carefully a 'scrubbed' female can be detected by a practiced eye.

The man took the five-pound keeper and placed it downwards, tail towards him, on the pegboard. He placed his right arm across the right claw. With his left hand he closed the left claw with a peg. The lobster retaliated and jettisoned two streams of fluid from under its large feelers. The man dodged the jets. The fluid contained a repellent, something like a skunk's, but not so foul smelling. Then the man pegged the right claw, gently tucked the tail under itself and placed the lobster in the appropriate basket.

The man proceeded down the string of marker buoys in the routine that was both his living and a flirtation with the eccentricities of the sea. He had, as best he could, set 400 pots in methodical order as befits the human mind. Nature in her whimsey will twist any grid system into a bent and irregular plan. The man and nature sparred, the one searching out the straightest sea bottom path, the other molding and changing things so that it did not conform easily to any order. It was not a co-operation between the two, but a condition set by the sea that had to be studied for its ways, as the ways of lobsters had to be studied through the catch in pots. But the man kept order, arranging pots in sets of 20, sometimes 10, sometimes 40, according to the bottom shape, and marking them for set and position in the run: A1, A2, A3, . . . A20; B1, B2, B3, . . . B20; C1, C2, . . .

Each year he started out with method, and each year he worked to maintain it. Storm damage he could accept as part of his loss. A modern problem was warp cut by propellers from power driven boats. Because of the loss of a buoy and the inherent trouble of finding the unmarked pot, he positioned the wafer buoy about 10 fathoms up from the pot and not 3 fathoms as is the custom. At low tide the wafer floats near the surface; not permanently way under and out of sight. He would start the season with 40 reserve buoys to replace those lost and by the end have none left.

On the fifth pot of the third run the buffeting seas slued Lilly B awkwardly into the rising warp. It fouled the prop. The man hove-to immediately, cursed, picked up the gaff, searched around in the water off the stern, found the warp and hauled it and the pot aboard. Waves slapped against the hull like impatient fingers tapping a desk. No amount of manipulating would free the line from under, so the man turned off the engine, opened a trap in the deck and reversed the shaft by hand. The warp would not come free and the man assumed it had looped around a blade. To save his gear he cut the warp close to the fouling to free it and bent the ends with a trawl knot, a single overhand knot at one end with the other backed into it for the length of the first knot. He checked the pot for the catch and cautiously felt for steerage. The propeller bit into the seas and the man was satisfied. If there was spaghetti-like warp wrapped around the shaft it would have to remain; he was not going overboard in these seas to inspect the fouling and, in any case, he was not one who enjoyed risks with inquisitive sharks.

The delay might have upset another man's fishing, but this man came back to his routine. Checking the face of his yellowing, scratched, Timex watch, and noting that 10:00 A.M. was approaching, he reached for the switch of the Apelco transmitter. The click was lost to the wind. But as the equipment warmed the background presence went forth from Lilly B in all directions. There were ears that had heard and there was one person who had been waiting, although she had no corresponding transmitter to respond. She noted the presence over the atmospherics and knew the hand that had flipped the switch. She had heard the power come into play, knew that her man was somewhere out there. That much warmed her. And the man knew it, too.

CHAPTER EIGHT

LILLIAN BLOOMER paused in her household work at that time in the morning that subconsciously comes through to her before ten. She unplugged the electric iron, switched on the S120 Hallicrafters tucked in the corner of the kitchen, and glanced at the clock on the wall.

She made a cup of coffee, increased the volume a fraction and took an easy chair in the living room. Automatically she reached for a pack of Salem cigarettes on a side table and went through the motions of smoking. She inhaled gently, not because it was her usual custom to be gentle, but of late she did so because of the tenderness there: the result from a cancer operation. She puffed and raised her head in thought. One could not know what she was thinking, whether it might be about radium treatments and the future, or whether her mind envisaged Lilly B and her husband working for their daily bread.

The radio crackled out extraneous atmospheric noises that scrambled any intelligent sound trying to reach home. Any other day in summer, when the weather is fine, seas smooth, 27.38 Kc/s would carry boat to boat voices without space for a commercial message. Then you might hear the man's voice on top coming over loud asking for those voices to give up a moment. The voices would continue, and the man would haul the next pot while he listened and waited for a break—patiently being impatient.

Right now it was mostly the atmospherics that were not letting up.

At precisely the hour the radio changed its breathing, and one could detect a presence over the ether. It was a presence without a voice, but one knew a voice would come in.

"Lilly B. Calling the Thelma."

Thelma is "Sonny" Mallowes' boat, and there was no Thelma out there as those concerned would know. There was a pause, and the sense of the seas off Monomoy Point came through. It permeated the living room where the man's wife sat. It swept in like a wave, bringing with it a far off distance of another world, one of movement in the oceans where a being was fencing with forces more powerful than his, where his skills would be diminuative against a chance unique capricious turn in wind and currents and counter reactions. The flowing in brought with it a suspense that said there was something to tell, something that was not supposed to be with the daily routine, but it was O.K. nevertheless. The appointed time always brought concern, an anxious moment, and it always would, as it should between husband and wife. The presence of the ether and the call eased the tension somewhat, for even without further message the linkage had accomplished its initial purpose.

As though the man was in the middle of a conversation, he continued to speak. He said his piece about the catch, the slowness of it all, that he would be in late—what with the weather and he being behind schedule.

Without pause and running his sentences together as if the sense was part of the whole he said, "I've got a blue if anybody wants it." And cut off.

This was not the regulation way for salutations and those in the area knew it, understood, accepted it as the ways of men when they have things on their mind other than operating to the rules.

The man plowed on earning his daily bread. Lillian Bloomer turned her attention to a music score on an electric organ in the living room. Tomorrow, at the local St. Christopher Episcopal church, she would play at a wedding.

CHAPTER NINE

LIFE'S ROUTINE for husband and wife moved on with an order suited to each temperament. They would meet again with mutual planning, the wife to listen for the presence that should come to her near 12:20 p.m. when her man would note the time in his trapping.

She, by then, would be along in her household chores prior to departing for part-time employment at the Silhouette Shop on Main Street in Chatham. She might, if the timing would permit, take the route by the sea to watch for Lilly B coming in. It reassured her before she attended to the needs of shoppers with their singular summer thoughts of hats and shorts and sandals, and an evening meal of lobsters.

The presence would accompany her to town and remain hidden with her during her store hours. It would not be read in her countenance, and the bearing of her rigid carriage would show no hint, but it would be somewhere there: of pots and buoys, and warp, and tossing, and hauling; mixed with merchandise and people-vacationers, for although she was not one for the sea she was not without knowledge. This was no jumbling of two environments but an intertwining that bound her to her man, as the wife of a coal miner, or the wife of a sweating worker at a steel furnace, or the wife of an airline pilot now that hijacking was here.

She felt strong in her man's accomplishments. It was a

strength that gave her daily confidence. His professionalism she accepted as a natural reward of marriage because she knew it had been handed down between the generations. Hers was not a marriage on a shaky future, fast bucks, ever changing jobs for the lack of a skill, or not knowing what life was about and the mystique of trying to find oneself in it, or the restless searching for an illusion, or travelling to some undefined utopia. Hers commenced with treading a measure at a dance and wondering at the reserve of the young man, and liking it. She knew, when he proposed, where she was going.

The phone rang and she stopped her organ practice. A call from Boston. Tomorrow, another check-up at the hospital. Another consultation.

The man was not aware of the call but he knew one was expected. The reflection was there, tucked in dormantly as he timed his pace to hauling pots.

He normally hauled at the rate of one every two minutes. That included the time to clear the end of a string and get to the head of the next. It was not beyond him to increase the pace under favorable weather to almost one a minute, that is, from pot to pot. But he would "break his neck" doing it. Two minutes to a pot gave him more time to think, more time to be careful with gear, and if he had repair work, he needed that time anyway.

Generally, he hauled 100 to 140 pots a day. On fine days, 140. On mediocre days, 120. On bad days, 100. On really blowy days, when he takes a big thrashing, his time erodes away and he can not reach even 100.

Today was one of those, a struggle to maintain an average of a pot in three minutes, what with the fouling of his warp, the replacing of two wafers, and repairing a broken pot slat. He thought that was enough.

The man knew, though, that the irritants to his ongoings were undramatic articles of fate that caused most trouble.

Two years ago he had had a calm day, could not have been calmer, and he had finished hauling a good quota of pots. He was off the southerly tip of Handkerchief Shoal, five miles out from the Point. The final pot of the string sank to the bottom and he aligned Lilly B for home. He left the wheel, as was his custom while he cleaned the mess from deck and catered to the comforts of his catch with a dousing for the last hour at sea, when his craft stopped planing.

The engine roared, but there was no headway. The clutch! He inspected the shaft through the deck hatch. It rotated, yet no amount of revving produced power in the water. The prop! And then he saw floating from the stern, and attached to Lilly B, the most awful mass of tough plastic film concocted from polymerization that a power skipper could dream of in a nightmare; a snarl impossible to free from topside.

By trial, the man discovered he could make slight headway if he kept the engine turning slowly. He headed for the Point and a sheltered shallow cove on the north-westerly shore near the old abandoned lighthouse. A slow hour passed before he reached the golden sands reflecting through the clear waters of that welcome bay he knew well from school days. The ghost of the man's grandfather must have stretched forth supporting hands, for this was the borning place of his ancestor. One might ponder what the old man would have thought, seeing his grandson in such a modern craft, built in Jonesport, Maine, where that state was second to Massachusetts in pioneering commercial lobstering. A boat without oars! What is the matter with you boy!

Lilly B drifted close to land in 6 feet of water and lazily eddied in the smoothness. The man took no chances and searched around for bait fish, pogies or squid that swingtail sharks follow into shore where blue fish have chased them. It was a caution brought about from hauling multilated buoys, scoured by teeth. And when it was wooden buoys and not synthetic ones, he collected shark's teeth that were embedded deep.

> "When the sands are all dry, he is gay as a lark,
> And will talk in contemptuous tones of the Shark:
> But, when the tide rises and sharks are around,
> His voice has a timid and tremulous sound."
>
> *(Lewis Carroll: Alice's Adventures in Wonderland X)*

This man was not foolish. Within halloing distance a lone surf fisherman was casting his line from shore and the trapper called to him to watch the waters, which he did. Then he stripped to the buff, took a knife and eased himself over the side and plunged down. Five times he plunged, and five times he tore away an armful of plastic before the prop was clear. Each armful he dumped into Lilly B, away from further damage, to himself, others, or even saints who might have cause to swear. And the man swore to the heavens as a good Methodist should. He was not one for going to church. He did not think it necessary for fishermen, farmers or miners to go regularly, but he enjoyed going occasionally. Perhaps after this, next Sunday, he thought he might go.

He had found the waters warm which was part recompense. When he returned to the safety of his craft he went straightway to the cabin for a towel as if this might be the usual pause in his work. He rubbed himself down under the warmth of the sun, and felt good.

That day there was extra concern for his lateness.

CHAPTER TEN

THIS DAY the man had checked the weather report from Boston on 24.06 Kc/s and had immediately called the Thelma at 12:20 pm. Lillian Bloomer knew the pitch. Her man would be late, quite late.

There is a scenic route in Chatham called "the Loop." At 1:30 pm the man's wife backed out of the garage onto Barcliff Avenue and, instead of driving to Main Street and the Silhouette Shop, took herself partly around the Loop, not for any scenic pleasure but to check, as far as she could from one high vantage point, the possibilities of sighting Lilly B coming out of the fog of the Sound.

It was not unusual for her to wait for her man to come in, somewhere around 1:30 pm, for re-assurance of his safety. She would never express it that way, but in the general run of town neighborliness she would collect some fish, or convey a message to her man, as if it were something to do before she took to selling garments during the afternoon. Today she knew she could not make that kind of connection.

So she turned towards town on Route 28, took a small arc around the rotary opposite the Mobil Station that was now buzzing with cars sopping up gasoline, swung off into Stage Harbor Road, hurried by Oyster Pond town beach filled with bathers and sun worshippers in the scantiest of coverings, across Cedar

Street, by Bridge Street without a glance, around the corner by the Old Mill Boat Yard into Champlain Road and into sight of Stage Harbor and the distant decommissioned Stage Harbor Light. Turning inland, she halted up the hill of that arm of Chatham called The Neck. There she turned the car around in a driveway, musing that the neighbor was not home and could not mind the trespassing, and slowly came about face with a substantial view, not only of the harbor and the finger-like stretch of Harding's Beach opposite, but also a section of the northerly part of Monomoy Island. Towards the Sound she saw nothing but mist. Beyond that were running seas.

She sat there, with the car pulled over onto the shoulder of the road and with the engine aimlessly purring. It was not the first time she had sat there waiting. She had sat there before with her time running out and leaving without knowing. . . . What does a wife think, waiting for her man? Does the mind go blank with the ticking minutes? The eyes searching and straining and smarting in the attempt to pick out abstract happenings on the horizon and turn them into Lilly B. What does a wife think, knowing that her man is troubled by the seas, and knowing that he held some kind of incident back from her, not for any kind of cheating on life but to hold her off from concern? Is the unspoken word the wrong act?

This day, too, the wife sat, and time ran out. At five minutes to the hour of two she drove off; duty—for selling garments to other women—was strong.

CHAPTER ELEVEN

THE ANSWER remained out there in the seas. The man knew where his wife had waited, when she had moved off, where she was headed. He knew she left without answer.

This man's trip may not have been absolutely necessary, but it was his way. He had to make the move. Lobsters were not plentiful. The demand was there, greater than the supply, the retail price ridiculous. And still there were those who would buy. "I need eight for a party, tonight. They must be 1¼ pounds apiece." Money flows in Chatham.

The demand for lobsters had not always been that way. When the Pilgrims landed they already knew about *Homarus,* but in Europe theirs was *Homarus gammarus;* the American lobster, *Homarus americanus,* was considered no better food. Lobsters piled eighteen inches high along the American shores after a storm. They were collected in carts and plowed into the fields for fertilizer. In colonial days they were trash food and household servants thought themselves fortunate if they ate them but twice a week.

Now they were precious cargo and expensive. This man took good care of his catch.

He did not have a holding tank like some others did, preferring to transport lobsters damp and cool in bushel baskets. A tank might give more problems, having to haul the catch out at the end of each trip, besides the expense of equipment to supply oxygen. Lobsters are pugnacious in their natural environment, finishing a fight to the death, topped off with cannibalism. When harvested they lack stamina.

With the last pot baited and sinking and the marker thrown out and following through the air, the man aligned Lilly B for home. Scurrying, he cleaned deck, swept the ocean bric-à-brac towards the stern, scooped the chaff into a dustpan and dumped the contents back into the sea. Then he sloshed water onto the deck and along the freeboard where there was the most debris from hauling pots, and scrubbed the planking where ocean life had cemented itself. He re-arranged the catch in the stern and doused them all over many times with buckets full of fresh bubbling water that he scooped out of the ocean. Each time he collected water he let it rush into the bucket as the seas swept by. It was a clean and refreshing operation after mauling the debris from the ocean floor. For the lobsters, this would be their last touch of the real environment; the salinity and chemicals peculiar to the place only a lobster can know about. No more love making. No more fighting. Theirs was now an imprisonment; shackled immobile by tiny wooden pegs, two apiece; each body wedged against a companion in fate—of cousins and aunts and brothers and sisters and mates, if they knew the distinctions; antennae waving slowly in attempts to orientate themselves and not knowing why they had lost their special senses now that they were air enveloped; twin eyestalks moving, inspecting this sudden lobster jam with passiveness; each giving forth bubbles from maxillae, like a new-born child. The movements of Lilly B, although she transmitted to them the forces of the sea that had once been theirs was of a different motion; that from the vessel was not to feel the varying pressures that had come to them deep down on the sea bottom, but they probably detected a kind of kinship, nevertheless.

The man flopped the brightly colored striped awning that served as a tarpaulin into the port wake that was forming from the bow, and pulled it back, heavy and dripping, over the baskets.

Now the man concentrated on a single purpose: to get to

his wharf safely without delay. Speed was precious; dallying increased the risks of keeping his catch fresh and alive. His mind was set on getting them weighed and checked in with Willard Nickerson, his lobster handler.

Lilly B took to the stirrups that dug into her and she tumbled and rolled over the roughness that was beneath her. Again it was an irksome speed for the man: irksome getting out to his fishing grounds; irksome making time in hauling; and the forces in no mood for letting up on the return trip. But the irksomeness was not grinding his nerves inwardly. He took to the task as part of his job, and because he had made the decision to go out he was not bugged by a master holding a whip. He had successfully accomplished the major part of his assigned work for this day, and he was a good hand in boisterous seas. With Lilly B on course he removed his working garments, gave a quick 360-degree scan and reached down into the cabin for his lunch pail. Within seconds he was back at the wheel checking again.

Happy satisfaction radiated from him as he munched his way into a submarine sandwich, and the tension of his jowls at each bite flexed a face that had a smooth skin, one that could have belonged to a man fifteen years his junior. Then he took a slug of hot coffee from the flask. It was a neat trick.

A man alone with his labors, when they are divided between a skill and a routine, is apt to be nursed into thoughts that tumble helter-skelter.

While he savored his contentedness, he reflected upon his decision to build Lilly B. It was in the fall of 1966 when he weighed the points of a boat ten to twelve years old, and how she gets loose and commences to leak, and that she either has to be fixed or sold. He sold Smallfry. But before he did, he nosied into boatyards on the Cape and along the indented Maine coast. He recalled that the yards made good boats but there was always something in their constructions that he did not like, and that all the while he was nosing around he kept hearing the name Bertram Frost, and that he ran the Frost Shipyard Company in Jonesport, but that he might be expensive.

A raft of immature common eiders slowly passed to port

and took his attention. He watched them rising and falling on the heaving waters, a tightly packed group of fifty ducks and not one interfering with the movements of a neighbor. He thought they might be resting because of the weather.

This reminded him of a small bird, a warbler, that had once flown onto the cuddy roof, pooped and bushed. Lilly B was no swaying branch, but she must have been the last asylum for tired wings. The bird took refuge with him for an hour.

Then the man flashed back to Frost and how he wanted his boat built. Frost said that he could build one that way: one that would keep its head up and stay up when it had been water-soaked; one with more sheer than usual—a more crooked bottom. He recalled looking at boats in the harbor there and saying, "One like those would run off to sea and throw me right down to the eyes." So they put their heads together and conceived Lilly B. He remembered how she was drafted, from the amalgam of two experiences, a boatbuilder and a fisherman, each matched with equal strength and New England disposition. They had a contract that was written, more to tie in their collective ideas, and to conform, more or less, to the day's business practices. When the boat was fitted out it finally cost near $15,000. He was proud of the deal. He was proud he could pay hard cash.

And then, when Lilly B was squared away, Frost asked to accompany him to the Cape. "Just for the ride," he had said. They took off into a snow storm, a gale of wind, and 15 degrees of temperature.

A droning overhead alerted him to a plane.

He wasn't alone, then. What poor soul of a pilot would be up there in this? Before the war he held a pilot's license; piloted for a hobby. He tried to enter the Air Force when war commenced, but his eyesight let him down. He settled for the Coast Guard. Perhaps it was just as well.

He passed the bell buoy. All morning long he had not seen or heard another sea craft, nor had he seen the faintest outline of land. Peculiar how some men fear being out of sight of land when in a small boat, like some swimmers who fear being out of reach of the bottom. There were those who even considered the sea their enemy, a thing to contest with, instead of respecting her and taking her favors when she had them to offer.

As the buoy disappeared it seemed to be drawn into the place that he had left. Solitary things buoys, rooted to the sea, leaning with the tides, swaying with the waves, creating strange eddies. They are things to note, and things to steer clear of.

He remembered the accident, too closely connected to dwell upon the details, where two men in an open boat and who knew these waters, hit a buoy. The proof of the hitting was the scraped paint discovered there. That day, during the search for the bodies, the whole of Chatham was drawn together especially close.

Like an old-time movie, past events flipped into memory-view without rhythm, without order. They came and went loosely. There was no control, and completely detached from the decisions the man was making at the wheel of Lilly B. It was as though he might be conducting a symphony orchestra in the auditorium of a conservatory of music where, in the next room, a band was blowing and thumping the blues.

"Jaws!" he said aloud. "I think I would like to see that film with Lil tonight in Hyannis, after I get that damn warp from off the shaft."

CHAPTER TWELVE

OVER THE RAIL of the drawbridge on Bridge Street men and young children dangled fishlines, and summer traffic rumbled the timbers and disturbed nobody. A great black-backed gull with a peculiar cast in one eye stood upon the ridge of the bait shanty, restlessly flying off and returning. This summer he had become a fixture, had learned the routine of the man, knew more about his timing and his catch and the cutting of bait and the leftovers than any of the people on the bridge would have suspected. When the bird stood awkward to the wind its feathers ruffled, and the sun was hazy enough to show him in muted gray. This dispassionate scavenger did not read the sun to a precise hour, but the light told him the afternoon was late and something was amiss with the man's habits.

"Sonny" drove by with his dog and noted the skiff at her mooring.

In barefeet, Tommy Ennis, proprietor of the Old Mill Boat Yard, climbed out of the slip and half expected to see Lilly B waiting at the end of the wharf for gas.

Willard Nickerson had not noticed, as yet, that his man had not arrived with lobsters, what with the rush of customers, the cutting of extra bass fillets, and keeping his business in order.

The town of Chatham was beating to the pulse of summer. The rhythm on beaches and town landings, in restaurants and stores, along byways and alleys, reverberated to the fringes of the vacationing mass. The rhythm was more from off-Cape bodies than from its own heartbeat, although its own kept in tune to make a living from it. Some of the pleasure-seeking pulses quickened because this was the annual let-loose. Some wondered and were envious of the town. These saw a glamor, could not imagine Chatham in a winter covering, a hurricane, or the supporting labors to keep the place going. There was temporary glorification in spending money, hard earned no doubt, but from under a different worldly roof.

Within the confines of a link-fence enclosing the Chatham Light, young men of the Coast Guard played volley-ball, exercising muscles and keeping from boredom.

Police in a cruiser checked the parking lot in front of the Light for violators on the new time limits, and a man with binoculars on the bluff squinted at the Pendleton wreck surrounded by breakers and asked anybody within hearing distance what it was. He could not see the other side of Monomoy Island where a lone fishing boat had passed Inward Point and was now skimming across the South'ard.

CHAPTER THIRTEEN

THAT EVENING, as the sun sank behind town, soft-edged shadows outlining the man's two shanties, crept across the crushed shell drive. A muskrat gingerly nosed its way under the drawbridge, now empty of dangling fishlines. A great black-backed gull, with a peculiar cast in one eye, flew off into the dusk. The place was much as it should be for that time in the evening except that on this day a short length of warp that had been casually thrown onto the bric-a-brac was missing.

When all was to himself, the great black-backed gull had pulled the warp out of the leavings and had run across the drive like a schoolboy with a prized treasure that he would covet for the rest of his days. He took it to the ragged grass by the quay edge and played tug-o-war, wrestling, twisting, poking, and pulling as though it might have been an exceedingly long thin eel. It did not react; and it was tough.

As the gull flew off, the warp began to slither down the quay face. It slithered into the water and disappeared, and the creeping shadows covered the spot to end the episode and a day's work.

. . . the sun sank behind town, . . .

A great black-backed gull . . .